THE WEATHER

SNOW

A Crabtree Roots Book

DOUGLAS BENDER

CRABTREE
Publishing Company
www.crabtreebooks.com

School-to-Home Support for Caregivers and Teachers

This book helps children grow by letting them practice reading. Here are a few guiding questions to help the reader with building his or her comprehension skills. Possible answers appear here in red.

Before Reading:

• What do I think this book is about?
  - *I think this book is about snow.*
  - *I think this book is about how snow is made.*

• What do I want to learn about this topic?
  - *I want to learn where snow comes from.*
  - *I want to learn what snow is made of.*

During Reading:

• I wonder why...
  - *I wonder why it doesn't snow all year round.*
  - *I wonder why snow melts.*

• What have I learned so far?
  - *I have learned that snow is made of water.*
  - *I have learned that snow forms in clouds.*

After Reading:

• What details did I learn about this topic?
  - *I have learned that we use snow shovels when it snows.*
  - *I have learned that a lot of snow creates a snowstorm.*

• Read the book again and look for the vocabulary words.
  - *I see the word **snowflakes** on page 6 and the word **snowstorm** on page 8. The other vocabulary words are found on page 14.*

Brr! This is snow.

Snow is made
of water.

Snow forms in **clouds**.

Snow comes down as **snowflakes**.

A lot of snow makes a **snowstorm**.

We use snow **shovels** when it snows.

People like to jump
and play in the snow!

Word List
Sight Words

a	it	this
and	jump	to
as	like	use
comes	made	we
down	makes	when
in	play	
is	snow	

Words to Know

clouds

shovels

snowflakes

snowstorm

41 Words

Brr! This is snow.

Snow is made of water.

Snow forms in **clouds**.

Snow comes down as **snowflakes**.

A lot of snow makes a **snowstorm**.

We use snow **shovels** when it snows.

People like to jump and play in the snow!

SNOW

Written by: Douglas Bender

Designed by: Rhea Wallace

Series Development: James Earley

Proofreader: Janine Deschenes

Educational Consultant: Marie Lemke M.Ed.

Photographs:
Shutterstock: Taiga: cover; Nick Starichenko: p. 1; Triff: p. 3, 14; Key_Keeper: p. 4; thebezz: p. 5, 14; Bobkov Evgenly: p. 6-7, 14; stoatphoto: p. 9, 14; Rossario: p. 11, 14; Serg Zastavkin: p. 12

Library and Archives Canada Cataloguing in Publication

Title: Snow / Douglas Bender.
Names: Bender, Douglas, 1992- author.
Description: Series statement: The weather forecast | "A Crabtree roots book".
Identifiers: Canadiana (print) 20210181354 | Canadiana (ebook) 20210181362 | ISBN 9781427159366 (hardcover) | ISBN 9781427159427 (softcover) | ISBN 9781427133731 (HTML) | ISBN 9781427134332 (EPUB) | ISBN 9781427159601 (read-along ebook)
Subjects: LCSH: Snow—Juvenile literature.
Classification: LCC QC926.37 .B462 2022 | DDC j551.57/84—dc23

Library of Congress Cataloging-in-Publication Data

Available at the Library of Congress

Crabtree Publishing Company

www.crabtreebooks.com   1-800-387-7650

Printed in the U.S.A./062021/CG20210401

Published in the United States
Crabtree Publishing
347 Fifth Avenue, Suite 1402-145
New York, NY, 10016

Published in Canada
Crabtree Publishing
616 Welland Ave.
St. Catharines, Ontario L2M 5V6